Bibliographic information published by the German National Library:

The German National Library lists this publication in the National Bibliography; detailed bibliographic data are available on the Internet at http://dnb.dnb.de .

Imprint:

Print and binding: Books on Demand GmbH, Norderstedt Germany
ISBN: 9783668965911

This book at GRIN:

https://www.grin.com/document/462115

Jagadish Jadhav

Compact Broadband Antenna Integrated with Filter

GRIN Verlag

Compact Broadband Antennas Integrated with Filters

JAGADISH BABURAO JADHAV

Department of Electronics & Telecommunication Engineering

R. C. Patel Institute of Technology, Shirpur

NOVEMBER 2019

Abstract

In recent years, there has been constant growth in using gigahertz frequencies for various applications. The state of the art of antenna integration with passive circuits techniques offering increased cut-off frequencies, open up new opportunities for integrating systems along with an antenna for either Transmit/Receive application. The work carried out in this seminar mainly deals with the survey of antenna structures which are compatible to integrate with passive circuits for microwave as well as milimeter/sub-milimeter wave applications. Lot of research in the field of passive receiver frontend design is being carried out. This work covers three aspects of passive receiver frontend design. The first is the analysis and design of new wideband high gain microstrip antennas, the second aspect is the design of compact high performance low-pass and band-pass filters and the third aspects is to integrate antenna and filter into single substrate. The trend of communication systems has imposed many restrictions on designers to have systems which should be light weight, robustness and have easy integration to other microwave circuits. In most of the communications systems the receiving antenna is followed with a band-pass filter. In microwave band this is normally a distributed filter and is implemented using the transmission line resonators. The transmission line filters are not compact, and in many applications where size is an issue may not provide the best solution. In the new active arrays, for example, it is desirable to integrate a complete front end with each individual antenna element. The area allocated to each cell, however, is determined by the array design and apertures efficiency requirements and is barely integrate the antenna and band pass filter in a single module, as is it more compact may also decrease the pre-filtering losses to improve the noise performance. The design of such system is a challenging task.

Contents

List of Figures

List of Tables

Abbreviations

MSA	Microstrip Antenna
BPF	Band Pass Filter
CAD	Computer Aided Design
dB	Decibel
DGS	Defected Ground Structures
DSS	Dumbbeii Shape Slot
EBG	Electromagnetic Band Gap
EM	Electromagnetic
FBW	Fractional Bandwidth
GHz	Gigahertz
HTS	High Temperature Superconductor
IEEE	Institute of Electrical and Electronics Engineers
LTCC	Low -Temperature Cofired Ceramics
MEMS	Microelectromechanic Systems
MIC	Microwave Integrated Circuits
MMIC	Monolithic Microwave Integrated Circuits
RF	Radio Frequency
SIR	Stepped Impedance Resonator
SSS	Spiral- Shaped Slot
TEM	Transverse Electromagnetic propagation Mode
TZ	Transmission Zero
UIR	Uniform Impedance Resonator
WLAN	Wireless Local Area Network
WiMAX	Worldwide Interoperability for Microwave Access
UWB	Ultra Wide Band

Chapter 1

Introduction

1.1 Antenna for Wireless Application

Definition: A metallic apparatus for sending and receiving electromagnetic waves. In other words the antenna is the transitional structure between free-space and a guiding device. It is used to transport electromagnetic energy from the transmitting source to the antenna, or from the antenna to the receiver. In the former case, we have a transmitting antenna and in the latter a receiving antenna. For wireless communication systems, the antenna is one of the most critical components. A good design of the antenna can relax system requirements and improve overall system performance [6].

1.2 Filters for Wireless Application

Electronic filters are circuits that have signal processing functions. i.e. they transform an input signal to obtain an output signal with the required characteristics. In the frequency domain filters are used to reject unwanted signal frequencies and to pass signals of desired frequencies. Filters are indispensable devices, in many systems and applications including wireless broadband, mobile, and satellite communications, radar, navigation, sensing and other systems. With the development of these systems, mostly induced by great commercial interests, limited electromagnetic spectrum has to be shared among more and more systems. Thus, there is an increasing demand for RF, microwave, and millimetre wave filters with more stringent requirements. These filters are employed in various systems to select or confine signals within specified spectral limits [15].

1.3 Design Consideration Parameters of MSA

The design of the microstrip antennas (MSA) is critical at higher frequencies due to a number of factors. For example, conventional microstrip antennas at lower frequencies use thin substrates and generally have narrow bandwidth performance. One method of improving the bandwidth is to use a thicker substrate. Therefore, as these substrates would be electrically thick at higher frequencies, one would expect a higher bandwidth, but at the same time the losses would increase and hence must be compensated. In addition, as the substrate thickness is increased, higher order surface-wave modes may propagate. The following characteristics must be considered while designing the microstrip antenna elements at higher frequencies.

- Substrate properties
- Resonant element length
- Resonant input resistance
- Bandwidth
- Losses due to surface waves
- Losses due to dielectric
- Dispersion effect

1.3.1 Substrate Properties

The most commonly used substrate materials are Unreinforced Poly Tetra Fluoro Ethylene (PTFE), Teflon, Reinforced PTFE, RT Duriod 5880, Fused Quartz, 96% Alumina etc. Electrical properties of these materials are summarized in the Table. The data are at X-band (8-12 GHz) and room temperature. However, the data are reasonably accurate for millimeter wave microstrip antenna design. The main reason for this is that, the dependence of dielectric constant on frequency is known to be low, especially for frequencies ranges of up to 110 GHz, where microstrip antennas are typically being used.

Material	Dielectric constant	Loss tangent
Unreinforced PTFE, Teflon	2.1	0.0004
Reinforced PTFE, RT-Duriod	2.2	0.0009
Fused Quartz	3.78	0.0001
96% Alumina	9.40	0.0010

TABLE 1.1: Electrical properties of commonly used substrate materials for MSAs

Substrate selection for a microstrip antenna design is guided by the following criterion:

- Effects of dispersion on the dielectric constant and loss tangent of the substrate;
- Magnitude of copper and dielectric losses;
- Possibility of surface wave excitation;
- Anisotropy, if any, in the substrate;
- Environmental concerns, such as the effects of temperature, humidity
- Cost

1.3.2 Resonant Element Length

One of the considerations in the design of a rectangular microstrip antenna is the length L (radius r in case of circular MSA) for desired resonant frequency. This length is a function of substrate thickness h and dielectric constant r, and in case of a RMSA element, a function of width W. Because the dielectric fills only part of the region surrounding the antenna element, the resonant length does not scale with dielectric constant as an antenna in a homogeneous medium would.

Figure Represents variation of resonant length for the first resonance of a printed rectangular microstrip patch element versus substrate thickness h for two different substrates. The patch width is W = 0.3λ (λ is the free space wavelength). The performance of microstrip antenna element deteriorates for substrate thickness approximately 0.1 0 due to inductive input impedance locus. This effect occurs for both probe type and microstrip line type feeds. This situation is undesirable in most cases, and so the use of patches

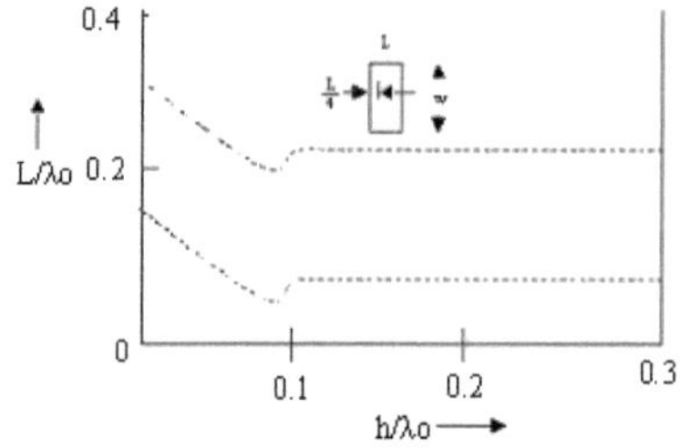

FIGURE 1.1: Resonant length versus height for (1) ϵ_r=2.55, W=0.3λ ; (2) ϵ_r=12.8, W=0.15 λ

on thick substrates may not be practical unless some way of countering this inductive trend, by using capacitive gap coupling from a feed line, for example, is used.

1.3.3 Resonant Input Resistance

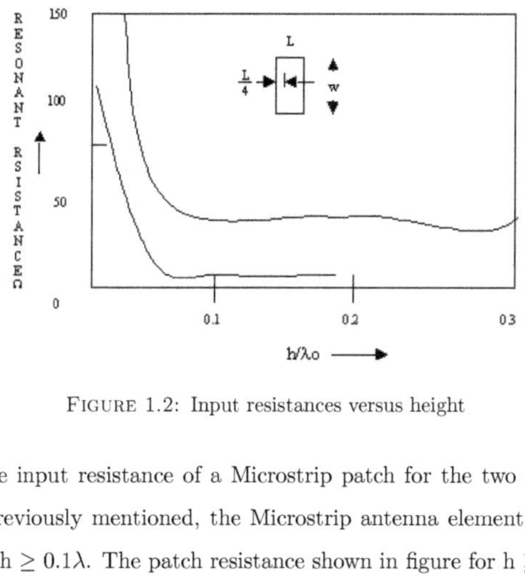

FIGURE 1.2: Input resistances versus height

Figure shows the input resistance of a Microstrip patch for the two substrates versus thickness. As previously mentioned, the Microstrip antenna element performance becomes worse for h $\geq$ 0.1λ. The patch resistance shown in figure for h $\geq$ 0.1λ is the real part of the input impedance for a patch length of 0.27λ.

1.3.4 Bandwidth

The impedance bandwidth of the MSA is defined as the frequency range over which it is matched with that of the feed line. This bandwidth is usually specified for voltage standing wave ratio (VSWR) less than 2. Sometimes for stringent applications, VSWR requirement is specified to be less than 1.5.

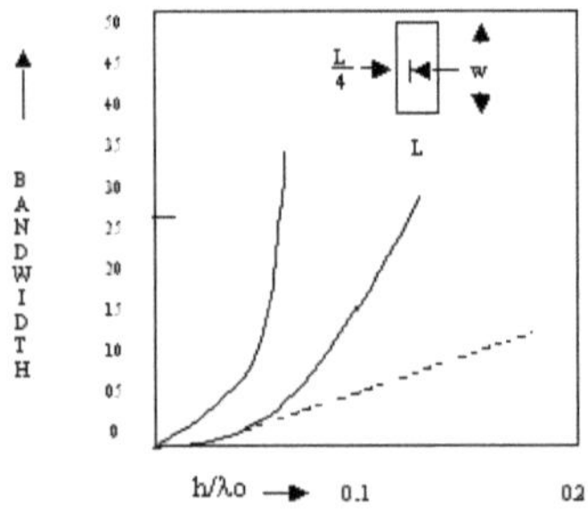

FIGURE 1.3: Bandwidth versus height

Figure shows the calculated bandwidth of a MSA versus substrate thickness for a PTFE substrate. The patch width is 0.3 0 and is fed by a probe at a point L/4 from the patch edge, although the feeding method or position does not affect the intrinsic patch bandwidth. The bandwidth increase rapidly with increasing substrate thickness, so that bandwidths of 10 to 20% can be obtained for substrate thickness in the range of h = 0.1λ to 0.2λ.

1.3.5 Losses due to Surface Waves

Two types of surface wave modes are possible, transverse magnetic (TM) and transverse electric (TE). For both types, the field components would vary sinusoidal with z in the substrate, and decrease exponentially in the z direction outside the substrate, where the various axes are shown in figure.

In general, the mode spectrum of a grounded dielectric slab contains both TE and TM modes, as shown in figure. The interval between the cut-off frequencies of two surface wave modes is given by

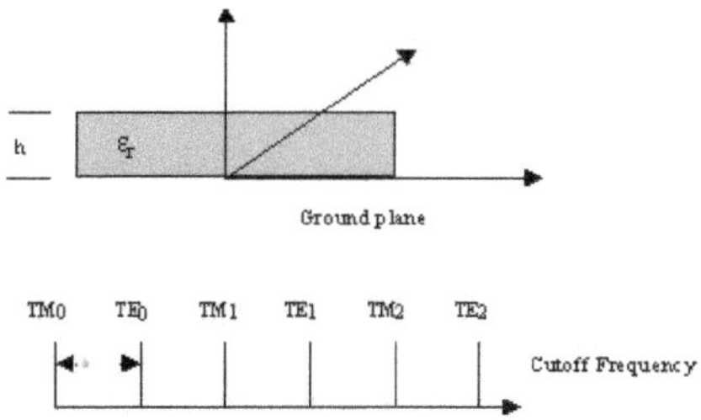

FIGURE 1.4: Dielectric slabs & its mode

F=$\frac{c}{4h\sqrt{\{\epsilon+1}}$

where c=$3\times10^{8}m/s$

As the TM0 mode has a zero cut-off frequency, this mode is always present in the substrate, regardless of the values of the substrate height and dielectric constant. For this mode, the y-component of the electric is strong, especially if the substrate is electrically thick or has a high dielectric constant. This characteristic affects the performance of both the single element and arrays. The higher order surface wave modes can be made non-propagating by choosing low values for h and r. Low values of h and r also help in reducing the magnitude of the TM0 mode to an acceptably low level. The price one pays for using low r is increased antenna size, while the price for using too low a value of h is possible reduction in antenna band width and radiation efficiency and thus a trade off is required.

1.3.6 Losses due to Dielectric

Power losses due to dielectric heating can be calculated by using the loss tangent and complex permittivity for the particular dielectric materials. Efficiency based on dielectric loss for a microstrip patch versus h is shown in figure.

1.3.7 Losses due to Dispersion

The dependence of the dielectric constant and loss tangent of the substrate on frequency leads to dispersion. For most dielectric materials of interest to the microstrip antenna designer, the dielectric constant decreases and the loss tangent increases with increasing

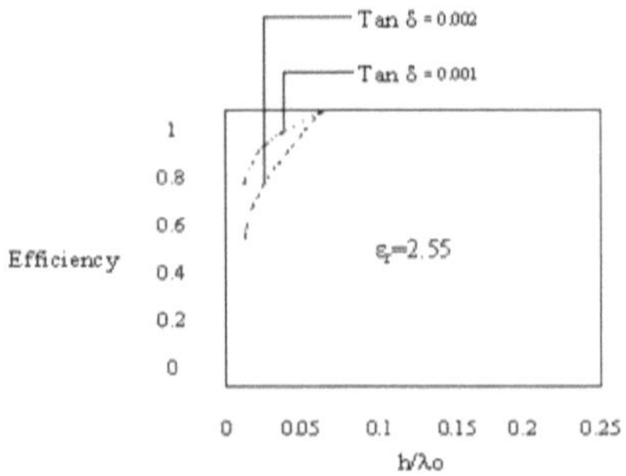

FIGURE 1.5: Efficiency versus height

frequency. The changes in the loss tangent are much more noticeable than changes in the dielectric constant, especially for frequencies up to 110 GHZ, which is the range of application of microstrip antenna.

1.4 Parameters of MSA

1.4.1 Effect of Width (W)

- Feed point location shifts towards the edge.
- Directivity increases, Area increases hence gain increases, Efficiency increases, Bandwidth increases.
- No change observed in the E-Plane but H-Plane decreases by 5%
- Resonant frequency slightly shifts downwards.
- If the width increases too much the higher order modes gets excited.

1.4.2 Effect of Dielectric Constant

- Good opportunity to reduce the size of the antenna.
- Resonant frequency shifts towards the upper side.
- Bandwidth and the gain decreases thus overall efficiency decreases.

- Feed point location shifts away from the edge. If dielectric constant decreases towards 1.0 the size of patch increases hence fringing fields and gain increases.

1.4.3 Effect of Height (h)

- Bandwidth increases directivity marginally increases.
- Resonant frequency shifts downwards.
- The height can be increased upto certain level after that efficiency decreases due to cross polarization level and surface waves.

1.4.4 Effect of Loss Tangent

- Substrate having low dielectric losses are expensive when loss tangent increases the dielectric loss increases which reduces the input impedance so feed point is shifted towards edge.
- Efficiency and gain decreases.
- No effect on resonant frequency and directivity.

1.4.5 Effect of Probe Diameter

- Generally SMA TNC and N-type connectors are used, as the probe diameter increases the probe inductance decreases hence input impedance decreases and resonant frequency slightly increases.
- No effect on gain and the directivity of the antenna.

1.4.6 Effect of Feed Point Location (x)

- If feed point is shifted from center towards edge impedance also increases, with the help of smith chart it is proved, resonant frequency slightly increases.

1.4.7 Effect of Radome

- Some times MSA is covered with a dielectric layer to protect them from the environment or some time radome material with some air gap is present with the MSA the following things are observed.
- Resonant frequency shifted downwards due to increase in effective dielectric constant.
- The bandwidth increases.

1.4.8 Effect of Ground Plane

Two types of ground plane are considered for MSA Finite and Infinite Plane. For the finite ground plane the resonant frequency is almost same but input impedance is slightly higher than that of infinite ground plane. Also for the finite ground plane the back lobes are present.

- MNM is the only method to analyze the infinite ground plane.
- The simulation time is reduced with infinite ground plane.

1.5 Introduction to the Problem

The trend of communication systems has imposed many restrictions on designers to have systems which should be light weight, robustness and have easy integration to other microwave circuits. In most of the communications systems the receiving antenna is followed with a bandpass filter. In microwave band this is normally a distributed filter and is implemented using the transmission line resonators. The transmission line filters are not compact, and in many applications where size is an issue may not provide the best solution. In the new active arrays, for example, it is desirable to integrate a complete front end with each individual antenna element. The area allocated to each cell, however, is determined by the array design and apertures efficiency requirements and is barely integrate the antenna and band pass filter in a single module, as is it more compact may also decrease the pre-filtering losses to improve the noise performance.

Such module may be viewed as filter, in which the last resonators and the load resistance have been replaced with resonant antenna element such as microstrip patch. The so called antenna filter devices have been attempted by the Naden, Coupez, Toustain, and Person in the past by using the multilayer technology and combination of high permittivity ceramic materials and foam substrates for the filter and antenna layers, respectively [1]. The foam layer helps in providing the high antenna efficiency and larger bandwidth, while ceramic material is used to reduce the size of the integrated filter and provide the low impedance level, which is required in the filter implementation. However, the multilayer composite implementation is not very convenient and required rather complicated binding procedure. Moreover, the specific design proposed in [1] fails to implement the filter parts in a single layer and uses an additional thin film microstrip and the connecting via holes to implemented parts of the filter. An earlier work by the Nawelaers and Capelle has addressed the subject in the context of the bandwidth enhancement, where it uses the resonance characteristics of the microstrip antenna and some external matching networks to achieve a second order Chebyshef response [2]. Microstrip patch antennas have drawn the attention of antenna engineers since the 1970s due to their attractive features of being low profile, compact size, light weight, and amenable to low-cost PCB (Printed Circuit Board) fabrication processes. However, patch elements are basically resonating at a single frequency, typically have less than 2% bandwidth [3] - [6], which is a major drawback in their utilization. Multiple techniques have been proposed to address the aforementioned bandwidth limitation. The simplest one is to use a thick substrate that potentially could enhance the bandwidth to 20% upon using 0.2 λ_0 - thick substrate as discussed in [7] by the Richards. However Pozar shows that would degrade the antenna efficiency due to undesired surface wave excitation [5]. Therefore, it is preferred to use either patches that are built with suspended substrates [8], or ones that are backed by cavities [9], [10], or even a combination of both [11] to achieve a wider operating bandwidth while minimizing surface wave excitation. Various approaches to Microstrip filter miniaturization are available through recent research. Some popular miniaturization methods are reported in literature include the use of slow wave resonators [12], multilayer filters [13], meandering [14], high dielectric-constant substrates, lumped element filters and dual mode resonators [16] - [18], [52] dual mode resonator based filters are naturally compact and generally offer the same fabrication simplicity as a regular Microstrip filter. However Hong and Sharman prove the highly compact dual-mode filter in recent research includes the open loop, and square loop

[17] structures. With wireless communication showing an unprecedented growth, filter design has seen more and more stringent requirements. They require low-loss, high power, highly selective filters with constrained physical size and economical viability in mass manufacturing. These demands led to advances in coaxial resonator, dielectric resonator, and superconducting filters, and also in methods of cost-reduction including computer aided alignment. Cellular radio handset manufacturing millions of extremely small, very low-cost filters, still with reasonably low loss and high selectivity. This has driven significant advances in integrated ceramic, surface, and bulk acoustic-wave active and filters using micromachined electromechanical systems. The slabline and microstrip filters were mainly designed for meeting terrestrial communication requirements. Due to their low insertion loss and higher power handling capabilities, combline and interdigital filters using slabline construction is widely used. Due to advances in material technology substrates with very good attenuation characteristics along with mechanical stability are now available. This has enabled filter designers to go for printed configuration where stripline filters were used earlier. Microstrip filters, with low loss substrates, have insertion loss comparable to slabline filters along with the added advantage of easy fabrication, compact size and easy integration with microwave circuits.

Chapter 2

Related Literature Review

The work carried out in this seminar mainly deals with the survey of antenna structures which are compatible to integrate with passive circuits for microwave as well as milimeter/sub-milimeter wave applications. The seminar briefly reviews the gigahertz applications. Thereafter the emphasis is on various possible antenna structures which are integrate with passive circuits. Such antenna structures are further compared in terms of their Gain, Bandwidth, Directivity, Coupling Efficiency and Compactness. The main focus of the seminar is towards the development of transreceiver front ends for gigahertz applications. In many wireless communication systems, the RF filters are usually placed right after the antenna. Since the size reduction and low profile structure are a trend in the circuit design, it is desired to integrate the bandpass filter and antenna in a single module, so called filtering antenna, with filtering and radiating functions simultaneously. Recently, several techniques have been proposed to reduce the size of wideband microstrip antennas [18] - [35]. In [18] - [26], various compact wideband microstrip antenna designs with slot cutting in the radiating patch are described. However, most of the slot-loading techniques shown by Girish kumar[22] , [23], such as U-slot patch antennas [18] - [21], V-slot loaded rectangular patch antenna [22] and E-shaped patch antennas [23] - [26], require a large antenna thickness over $0.06\lambda_0$. The large thickness of these slot-loaded patch antennas is mainly due to their thick air substrate. As another type of wideband single-layer single-patch antenna, the L-probe fed patch antenna proposed in [27], [48], [49], [28] by the Luke, and Lee exhibit a wide impedance bandwidth over 30 percent. However, as mentioned in [20], substrate thickness of the L-probe fed patch antenna will always be larger than $0.08\lambda_0$. The miniaturization of wideband patch antennas has also been accomplished by using shorting posts or shorting walls [29] - [31]. However, these shorted patch antennas usually suffer from the disadvantage of poor gain and degradation in the radiation patterns. The stacking of multiple resonators on different layers of dielectric substrates is also a practical approach for broadband operation [50], [51], but the total antenna height will be very large for lower operating frequencies [32], [33]. The chip-resistor loading technique shown by Wong is another promising technique to achieve wide impedance bandwidth and small antenna size [34] - [37]. However, due to the introduced ohmic loss of the chip-resistor loading, the antenna gain will be decreased significantly. However, to date, there has been relatively little research conducted on an efficient integration between the filter and antenna with simple fabrication and good circuit behavior. In traditional design, the filter and the antenna are designed individually, with the common ports characteristic impedance Z_0, and then connected directly. The direct connection of the filter and antenna usually causes an impedance mismatch, which may deteriorate the filters performance (especially near the band edges) and increase the insertion loss of the circuitry. To avoid this, extra matching network should be implemented in between these two components [1]. Several investigations have been focused on adding the radiating or filtering function into an antenna or filter [41], [42]. In [41], Froppier shown the work of metal posts were inserted into a horn antenna, which can generate the filtering function, and in [42], the way is to create coupled cavities into a leaky wave antenna so as to generate filtering performance on the antenna. While these proposed circuits possess the characteristics of filtering and radiating, it should be noted that they were designed without using a systematic approach not considering much on the filters or antennas specifications. Various approaches for integrating the filter and antenna into a single microwave device have been discussed in[43] by the Yang, Caloz and Wu. For size reduction, a pre-designed bandpass filter with suitable configuration was directly inserted into the feed position of a patch antenna. By using an extra impedance transformation structure in between the filter and the antenna, the bandpass filter can be integrated properly with the antenna over the required bandwidth [43]. However, the transition structure needs additional circuit area, and the designs did not have good filter characteristics over the frequency range. Recently, several filtering antennas designed following the synthesis process of the bandpass filter have been presented [1], [44]. In these designs, the last resonator and the load impedance of the bandpass filter were substituted by an antenna that exhibited a series or parallel RLC equivalent circuit. Although they have been done based on the co-design approach, most of them did not show good filter performance, especially the band-edge selectivity and stopband suppression. This is due to the lack of the extraction of the antennas equivalent circuit over a suitable bandwidth. Only that at the center frequency was extracted and used in the filter synthesis. Moreover, the antenna gain versus frequency, which is an important characteristic of the filtering antenna, was not examined in these studies. Although in [44] Tamijani shows, the frequency response of a factor named *ideal matching loss* was considered, it did not take into account the circuit and antenna losses and thus missed the power transmission characteristic of the filtering antenna. The main disadvantages of microstrip antennas are low gain, narrow bandwidth, and poor efficiency especially for antennas built using high dielectric substrate for easy integration with MIC RF front-end circuitry. Although integration of patch antennas with filters can be used to improve bandwidth and gain [53], the main purposes of such an integration is the reduction of size of the microwave front-end [45], [55], suppression of higher order antennas resonances [56] and the creation of wideband antennas with narrow band interferer rejection characteristics, for which band-stop filters are used [57]. As a result of integration, suppression of the first and the second spurious harmonics of antenna have been achieved. The integration was used to suppress the antennas peaks that are out of the passbands of the filters and to separate physically bands of antenna. Improvement in performance of the antenna and the physical separation of channels has been achieved as a result of integration. A rectangular microstrip patch antenna is a basic configuration of a microstrip antenna. The main disadvantages of this configuration are narrow band and spurious harmonics. Suppression of spurious harmonics of antenna can be achieved by, for example, etching a U-slot in the antennas patch [60], or by designing special feeding technique using proximity coupling [61]. The most obvious solution to the spurious harmonics problem is an integration of the antenna with a lowpass or bandpass filter [44]. One of the main challenges of such an integration using a single layer microstrip is that for microstrip patch antennas the most preferable type of substrate is a thick substrate with a small dielectric constant. Using this substrate, better efficiency and larger bandwidth of the antenna can be achieved. For microstrip filter, on the contrary, thin substrates with high dielectric constant are preferable as employment of this type of substrate reduces radiation losses and size of filters. To address this issue composite substrates and multilayer solutions can be used [62]. Another challenge is the size of the integrated antenna filters. The necessity of using impedance matching in the feeding of antenna can increase the size of the circuit substantially [56]. The integrated antenna-filter is a good solution for both of these challenges as compact pseudo-interdigital bandpass filter designed using low dielectric constant substrate will be used as a replacement of the impedance transformer employed as a feeding line of microstrip patch antenna. The length of the patch is approximately equal to half the wavelength.Microstrip filters are themselves already small in size compared with other filters such as waveguide filters. Nevertheless, for some applications where the size reduction is of primary importance, smaller microstrip filters are desirable, even though reducing the size of a filter generally leads to increased dissipation losses in a given material and hence reduced performance. Miniaturization of microstrip filters may be achieved by using high dielectric constant substrates or lumped elements, but very often for specified substrates, a change in the geometry of filters is required and therefore numerous new filter configurations become possible which include some novel concepts, methodologies, and designs for compact filters and filter miniaturization. The new types of filters include ladder line filters, pseudointerdigital line filters, compact open-loop and hairpin resonator filters, slow-wave resonator filters, miniaturized dual-mode filters, multilayer filters, lumped-element filters, and filters using high dielectric constant substrates. In general, the size of a microwave filter is proportional to the guided wavelength at which it operates [61]. The author Marjan Mokhtaari in [62] creates new possibilities of single or double notches while maintaining the circuit dimensions of the original UWB filter. Several prototype measurements validate the design approach and the additional circuitry for notch creation. Microstrip and stripline filters are based either on soft substrates or ceramic substrates with multi-layer thick film technologies. Microstrip transmission lines are broadband in frequency, compact and light in weight. Also they are of low cost since they can be easily adapted to hybrid and monolithic integrated circuit technologies. New design methodologies for microstrip filters are focussed on resonator geometry to produce miniaturized planner open-loop resonator filters, coupled-loop resonator filters, slow-wave-open-loop resonator filters, and loop filters in order to improve the out-of-band rejection. These filters are used to design compact duplexers in a specific band [52]. This size however scales with wavelength, which reduces there applicability for handheld devices below 2GHz. Most of the rapid advance in microstrip antennas and arrays took place in 1980s. Firstly this were driven by militry and space applications. Then this rapidely growing in commercial sector. Specifications for defence and space application antennas typically emphasize maximum performance with little constairnt on the cost. On the other hand commerencial application demands low cost components, often at the expense of reduced electriacal performance. Thus microstrip antennas for commerencial systems require low cost materials, simple and inexpensive fabrication techniques. Table shows the list of antennas used in commercial wireless systems.

System	**Antenna**
Cellular /PCS handset	Monopole, Sleeve Dipole, Folded F or Patch
Cellular / PCS base stations	Dipole or patch arrays
GPS Receiver	Microstrip Patch, Bifilar Helix
Direct Broadcast Satellite Subscriber	Reflector, slot array
WLAN	Monopole, Sleeve dipole, folded-F, or Patch
L M D S subscriber	Reflector, Patch array
Wireless local loop	Patch element or array, Monopole, Sleeve dipole
Point to point radio	Reflector, horn, or Patch array
Very small Aperture satellite radio	Reflector or patch array
RFID	Loop, folded-F, or monopole

TABLE 2.1: Antennas Used in Commercial Wireless Systems

Chapter 3

Parametric Study of MSA

3.1 Design Consideration Parameters of MSA

The design of the microstrip antennas (MSA) is critical at higher frequencies due to a number of factors. For example, conventional microstrip antennas at lower frequencies use thin substrates and generally have narrow bandwidth performance. One method of improving the bandwidth is to use a thicker substrate. Therefore, as these substrates would be electrically thick at higher frequencies, one would expect a higher bandwidth, but at the same time the losses would increase and hence must be compensated. In addition, as the substrate thickness is increased, higher order surface-wave modes may propagate. The following characteristics must be considered while designing the microstrip antenna elements at higher frequencies.

- Substrate properties
- Resonant element length
- Resonant input resistance
- Bandwidth
- Losses due to surface waves
- Losses due to dielectric
- Dispersion effect

3.1.1 Substrate Properties

The most commonly used substrate materials are Unreinforced Poly Tetra Fluoro Ethylene (PTFE), Teflon, Reinforced PTFE, RT Duriod 5880, Fused Quartz, 96% Alumina etc. Electrical properties of these materials are summarized in the Table.

Material	Dielectric constant	Loss tangent
Unreinforced PTFE, Teflon	2.1	0.0004
Reinforced PTFE, RT-Duriod	2.2	0.0009
Fused Quartz	3.78	0.0001
96% Alumina	9.40	0.0010

TABLE 3.1: Electrical properties of commonly used substrate materials for MSAs

The data are at X-band (8-12 GHz) and room temperature. However, the data are reasonably accurate for millimeter wave microstrip antenna design. The main reason for this is that, the dependence of dielectric constant on frequency is known to be low, especially for frequencies ranges of up to 110 GHz, where microstrip antennas are typically being used. Substrate selection for a microstrip antenna design is guided by the following criterion:

- Effects of dispersion on the dielectric constant and loss tangent of the substrate;
- Magnitude of copper and dielectric losses;
- Possibility of surface wave excitation;
- Anisotropy, if any, in the substrate;
- Environmental concerns, such as the effects of temperature, humidity
- Cost

3.1.2 Resonant Element Length

One of the considerations in the design of a rectangular microstrip antenna is the length L (radius r in case of circular MSA) for desired resonant frequency. This length is a function of substrate thickness h and dielectric constant r, and in case of a RMSA element, a function of width W. Because the dielectric fills only part of the region surrounding the antenna element, the resonant length does not scale with dielectric constant as an antenna in a homogeneous medium would.

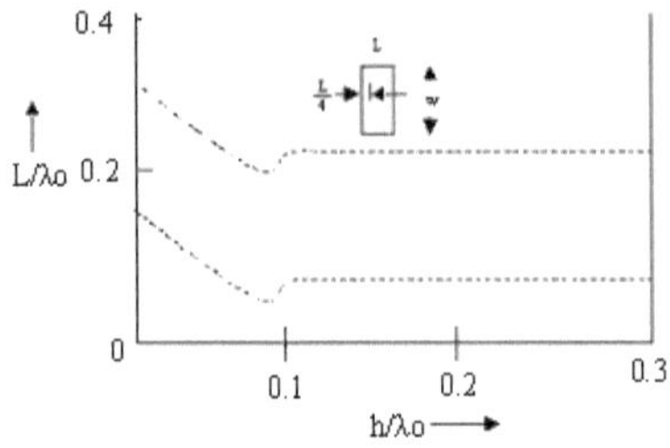

FIGURE 3.1: Resonant length versus height for (1) ϵ_r=2.55, W=0.3λ ; (2) ϵ_r=12.8, W=0.15 λ

Figure Represents variation of resonant length for the first resonance of a printed rectangular microstrip patch element versus substrate thickness h for two different substrates. The patch width is W = 0.3λ (λ is the free space wavelength). The performance of microstrip antenna element deteriorates for substrate thickness approximately 0.1 0 due to inductive input impedance locus. This effect occurs for both probe type and microstrip line type feeds. This situation is undesirable in most cases, and so the use of patches on thick substrates may not be practical unless some way of countering this inductive trend, by using capacitive gap coupling from a feed line, for example, is used.

3.1.3 Resonant Input Resistance

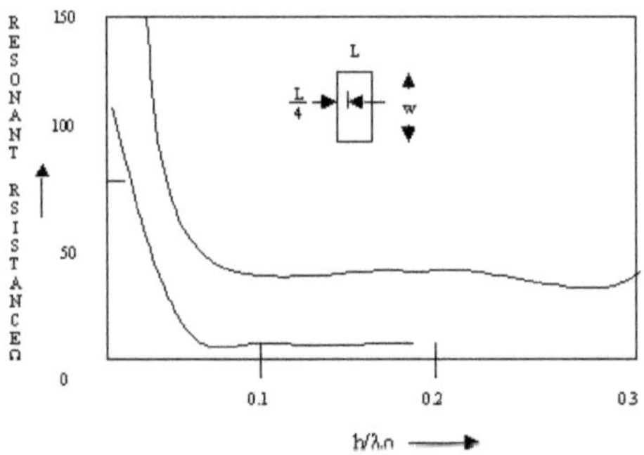

FIGURE 3.2: Input resistances versus height

Figure shows the input resistance of a Microstrip patch for the two substrates versus thickness. As previously mentioned, the Microstrip antenna element performance becomes worse for h $\geq 0.1\lambda$. The patch resistance shown in figure for h $\geq 0.1\lambda$ is the real part of the input impedance for a patch length of 0.27λ.

3.1.4 Bandwidth

The impedance bandwidth of the MSA is defined as the frequency range over which it is matched with that of the feed line. This bandwidth is usually specified for voltage standing wave ratio (VSWR) less than 2. Sometimes for stringent applications, VSWR requirement is specified to be less than 1.5.

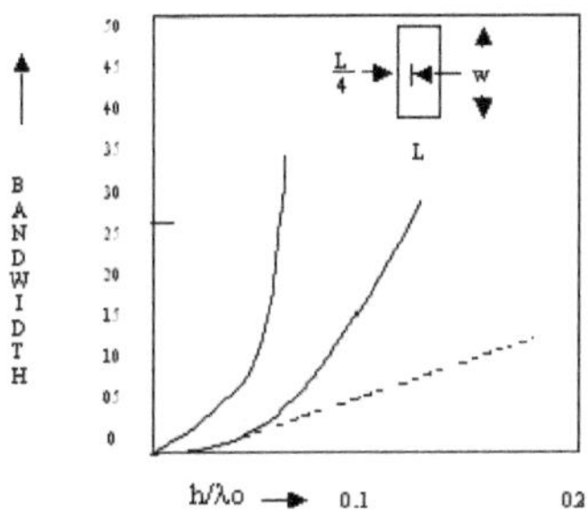

FIGURE 3.3: Bandwidth versus height

Figure shows the calculated bandwidth of a MSA versus substrate thickness for a PTFE substrate. The patch width is 0.3 0 and is fed by a probe at a point L/4 from the patch edge, although the feeding method or position does not affect the intrinsic patch bandwidth. The bandwidth increase rapidly with increasing substrate thickness, so that bandwidths of 10 to 20% can be obtained for substrate thickness in the range of h = 0.1λ to 0.2λ.

3.1.5 Losses due to Surface Waves

Two types of surface wave modes are possible, transverse magnetic (TM) and transverse electric (TE). For both types, the field components would vary sinusoidal with z in the substrate, and decrease exponentially in the z direction outside the substrate, where the various axes are shown in figure.

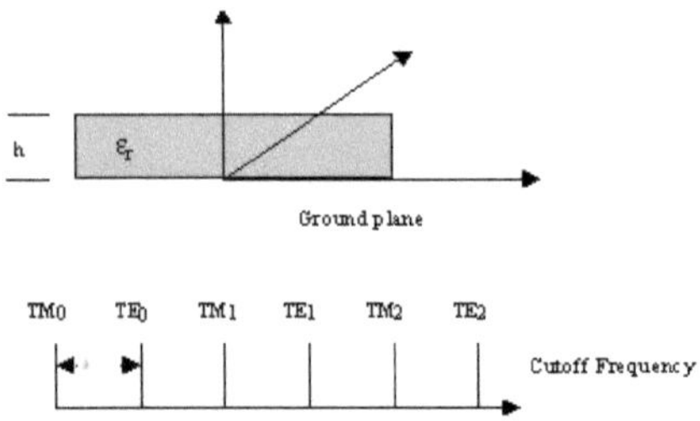

FIGURE 3.4: Dielectric slabs & its mode

In general, the mode spectrum of a grounded dielectric slab contains both TE and TM modes, as shown in figure. The interval between the cut-off frequencies of two surface wave modes is given by

$$F=\frac{c}{4h\sqrt{\{\epsilon+1}}$$

where c=$3\times10^{8}m/s$

As the TM0 mode has a zero cut-off frequency, this mode is always present in the substrate, regardless of the values of the substrate height and dielectric constant. For this mode, the y-component of the electric is strong, especially if the substrate is electrically thick or has a high dielectric constant. This characteristic affects the performance of both the single element and arrays. The higher order surface wave modes can be made non-propagating by choosing low values for h and r. Low values of h and r also help in reducing the magnitude of the TM0 mode to an acceptably low level. The price one pays for using low r is increased antenna size, while the price for using too low a value of h is possible reduction in antenna band width and radiation efficiency and thus a trade off is required.

3.1.6 Losses due to Dielectric

Power losses due to dielectric heating can be calculated by using the loss tangent and complex permittivity for the particular dielectric materials. Efficiency based on dielectric loss for a microstrip patch versus h is shown in figure.

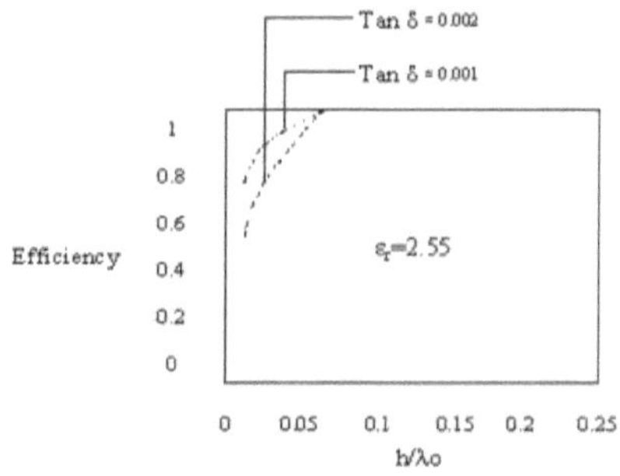

FIGURE 3.5: Efficiency versus height

3.1.7 Losses due to Dispersion

The dependence of the dielectric constant and loss tangent of the substrate on frequency leads to dispersion. For most dielectric materials of interest to the microstrip antenna designer, the dielectric constant decreases and the loss tangent increases with increasing frequency. The changes in the loss tangent are much more noticeable than changes in the dielectric constant, especially for frequencies up to 110 GHZ, which is the range of application of microstrip antenna.

3.2 Parameters of MSA

3.2.1 Effect of Width (W)

- Feed point location shifts towards the edge.
- Directivity increases, Area increases hence gain increases, Efficiency increases, Bandwidth increases.
- No change observed in the E-Plane but H-Plane decreases by 5%
- Resonant frequency slightly shifts downwards.
- If the width increases too much the higher order modes gets excited.

3.2.2 Effect of Dielectric Constant

- Good opportunity to reduce the size of the antenna.
- Resonant frequency shifts towards the upper side.
- Bandwidth and the gain decreases thus overall efficiency decreases.
- Feed point location shifts away from the edge. If dielectric constant decreases towards 1.0 the size of patch increases hence fringing fields and gain increases.

3.2.3 Effect of Height (h)

- Bandwidth increases directivity marginally increases.
- Resonant frequency shifts downwards.
- The height can be increased upto certain level after that efficiency decreases due to cross polarization level and surface waves.

3.2.4 Effect of Loss Tangent

- Substrate having low dielectric losses are expensive when loss tangent increases the dielectric loss increases which reduces the input impedance so feed point is shifted towards edge.
- Efficiency and gain decreases.
- No effect on resonant frequency and directivity.

3.2.5 Effect of Probe Diameter

- Generally SMA TNC and N-type connectors are used, as the probe diameter increases the probe inductance decreases hence input impedance decreases and resonant frequency slightly increases.
- No effect on gain and the directivity of the antenna.

3.2.6 Effect of Feed Point Location (x)

- If feed point is shifted from center towards edge impedance also increases, with the help of smith chart it is proved, resonant frequency slightly increases.

3.2.7 Effect of Radome

- Some times MSA is covered with a dielectric layer to protect them from the environment or some time radome material with some air gap is present with the MSA the following things are observed.
- Resonant frequency shifted downwards due to increase in effective dielectric constant.
- The bandwidth increases.

3.2.8 Effect of Ground Plane

Two types of ground plane are considered for MSA Finite and Infinite Plane. For the finite ground plane the resonant frequency is almost same but input impedance is slightly higher than that of infinite ground plane. Also for the finite ground plane the back lobes are present.

- MNM is the only method to analyze the infinite ground plane.
- The simulation time is reduced with infinite ground plane.

Chapter 4

Remarks and Conclusions

4.1 Remarks

Considering the recent widespread interest in antenna design and filters for Wireless communication, it is decided to concentrate research activity orientation towards the method based on method of moments approach. The use of these methods will allows us to obtain important results regarding analysis and design of antenna with filters. In proposed research work, we will investigate the design of compact broadband antenna as well as integrate it with filters for passive wireless transceiver frond end. The main disadvantages of compact antennas are low gain, narrow bandwidth, and poor efficiency especially for antennas built using high dielectric substrate for easy integration with MIC RF front-end circuitry. Although integration of antennas with filters can be used to improve bandwidth and gain, the main purposes of such integration is the reduction of size of the microwave front-end, suppression of higher order antennas resonances and the creation of wideband antennas with narrow band interferer rejection characteristics, for which band-stop filters are used. One of the main challenges of such an integration using a single layer microstrip is that for microstrip patch antennas the most preferable type of substrate is a thick substrate with a small dielectric constant. Using this substrate, better efficiency and larger bandwidth of the antenna can be achieved. For microstrip filters, on the contrary, thin substrates with high dielectric constant are preferred as use of this type of substrate reduces radiation losses and size of filters. To address this issue composite substrates and multilayer solutions can be used. Another challenge is the size of the integrated antenna filters. The necessity of using impedance matching in the feeding of antenna can increase the size of the circuit substantially. The integration of the filter & the antenna into just one module leads to a reduction of size and weight in the RF front end design, this integration can be seen as a contribution to greater compactness of the RF frontend. The compact low loss integration techniques enables highly efficient RF front ends with reduced sized & higher signal sensitivity.

4.2 Conclusions

- Narrow impedance BW is the most serious disadvantage of MSA.
- In the case of low frequency range such as Advanced Mobill Phone System[AMPS] (824-894 MHz), GSM (890-960 MHz), Digital Communication System [DCS] (1710- 1880 MHz) band, the size of MSA's are usually too large to be installed into a mini system.
- The problem of achieving a wide impedance BW for a compact MSA is becoming an important issue in modern MSA design.
- Low profile configurations (compact), wide impedance BW and high gain are important design considerations.
- For microstrip filter, thin substrates with high dielectric constant are preferred as use of this type of substrate reduces radiation losses and size of filters
- The antenna and filter integrated within the module can be independently designed at the beginning of the process and then incorporated with the proper design of the coupling between resonators.

Bibliography

[1] T. Le. Nadan, J. P. Coupez, S. Toutain, C. Person, "Optimization and miniaturization of a filter/antenna multifunction module using a composite ceramic foam substrate", *IEEE MIT-S Digest*, pp. 219-222. 1999.

[2] H. An. Nawelaers, A Van De Capelle,"A new approach of broadband Microstrip antenna design", *IEEE AP-S Digest*, pp 475-478. 1992.

[3] Y. T. Lo, D. Solomon, and W. F. Richards, "Theory and experiment on microstrip antennas", *IEEE Trans. Antennas Propag.*, vol. 27, pp. 137-145, 1979.

[4] J. R. James and P. S. Hall, *Handbook of Microstrip Antennas: IEEE Electromagnetic Waves Series.*, London, U.K.: Inst. Elect. Eng., 1989.

[5] D. M. Pozar, "Microstrip antennas", *Proc. IEEE*, vol. 80, pp. 79-91, Jan. 1992.

[6] C. A. Balanis, *Antenna Theory: Analysis and Design*, 3rd ed. Hoboken, NJ: Wiley, 2005.

[7] E. Chang, S. A. Long, and W. F. Richards, "An experimental investigation of electrically thick rectangular microstrip antennas", *IEEE Trans. Antennas Propag.*, vol. 34, pp. 767-772, Jun. 1986.

[8] M. Shahabadi, D. Busuioc, A. Borji, and S. Safavi-Naeini, "Low-cost, high-efficiency quasi-planar array of waveguide-fed circularly polarized microstrip antennas", *IEEE Trans. Antennas Propag.*, vol. 53, pp. 2036-2043, Jun. 2005.

[9] N. C. Karmakar, "Investigations into a cavity-backed circular-patch antenna", *IEEE Trans. Antennas Propag.*, vol. 50, pp. 1706-1715, Dec. 2002.

[10] F. Zavosh and J. T. Aberle, "Infinite phased-arrays of cavity-backed patches", *IEEE Trans. Antennas Propag.*, vol. 42, pp. 390-398, Mar. 1994.

[11] S. Yang and A. E. Fathy, "Cavity-backed patch shared aperture antenna array approach for mobile DBS applications", *in Proc. Antennas Propag. Society Int. Symp.*, pp. 3959-3962, 2006.

[12] J. S. Hong and M. J. Lancaster, "Theory and experiment of novel Microstrip slow-wave open-loop resonator filters.", *IEEE Trans. Microw. Theory Tech.*, vol. 45, no.12 pp. 2358-2365, Dec. 1997.

[13] J. S. Hong and M. J. Lancaster, "Aperture-coupled microstrip open-loop resonators and their applications to the design of novel Microstrip bandpass filters", *IEEE Trans. Microw. Theory Tech.*, vol. 47, no.9, pp. 1848-1855, Sep. 1999.

[14] S. M. Wang, C. H. Chi, M. Y. Hsieh, and C.-Y. Chang, "Miniaturized spurious passband suppression microstrip filter using meandered parallel coupled lines", *IEEE Trans. Microw. Theory Tech.*, vol. 53, no.2, pp. 747-753 Feb. 2005.

[15] J. S. Hong, and M. J. Lancaster, *Microstrip Filters for RF/Microwave Applications.*, New York: Wiley, 2001.

[16] J. S. Hong, and H. Sharman, and Y. H. Chun, "Dual-mode microstrip open-loop resonators and filters", *IEEE Trans. Microw. Theory Tech.*, vol. 55, no. 8, pp. 1764-1770. Aug. 2007.

[17] L. H. Hseih and K. Chang, "Compact, low insertion-loss, sharp-rejection, and wideband microstrip bandpass filters.", *IEEE Trans. Microw. Theory Tech.*, vol. 51, no. 4, pp. 1241-1246 Apr. 2003.

[18] S. Weigand, G. H. Huff, K. H. Pan, and J. T. Bernhard, "Analysis and design of broadband single-layer rectangular U-slot microstrip patch antennas", *IEEE Trans. Antennas Propag.*, vol. 51, pp. 457-468, Mar. 2003.

[19] T. Huynh and K. F. Lee, "Single-layer single-patch wideband microstrip antenna.", *Electron. Lett.*,vol. 31, no. 6, pp. 1310-1312, Aug. 1995.

[20] R. Chair, C. Mak, K. Lee, K. Luk, and A. A. Kishk, "Miniature wideband half U-slot and half E-shaped patch antennas", *IEEE Trans. Antennas Propag.*, vol. 53, pp. 2645-2652, Aug. 2005.

[21] B. Ooi,"A double-II stub proximity feed U-slot patch antenna", *IEEE Trans. Antennas Propag.*,vol. 52, pp. 2491-2496, Sep. 2004.

[22] A. A. Deshmukh and G. Kumar, "Broadband compact V-slot loaded RMSAs", *Electron. Lett.*,vol. 42, no. 17, pp. 951-952, Aug. 2006.

[23] A. A. Deshmukh and G. Kumar, "Compact broadband U-slot loaded rectangular microstrip antennas", *Microw. Opt. Technol. Lett.*, vol. 46, no. 6, pp. 556-559, Sep. 2005.

[24] Y. Ge, K. P. Esselle, and T. S. Bird, "E-shaped patch antennas for highs speed wireless networks", *IEEE Trans. Antennas Propag.*, vol. 52, pp. 3213-3219, Dec. 2004.

[25] B. Ooi and M. Leong,"Novel design of broad-band stacked patch antenna", *IEEE Trans. Antennas Propag.*, vol. 50, pp. 1391-1395, Oct. 2002.

[26] F. Yang, X. Zhang, X. Ye, and Y. Rahmat-Samii, "Wide-band E-shaped patch antennas for wireless communications", *IEEE Trans. Antennas Propag.*, vol. 49, pp. 1094-1100, Jul. 2001.

[27] C. L. Mak, K. M. Luk, and K. F. Lee, "Microstrip line-fed L-strip patch antenna", *Proc. Inst. Elect. Eng. Microw. Antennas Propag.*, vol. 146, no. 4, pp. 282-284, Aug. 1999.

[28] Y. X. Guo, K. M. Luk, and K. F. Lee, "L-probe proximity-fed annular ring microstrip antennas", *IEEE Trans. Antennas Propag.*, vol. 49, pp. 19-21, Jan. 2001.

[29] Y. J. Wang, C. K. Lee, and W. J. Koh, "Single-patch and single-layer square microstrip antenna with 67.5 percent bandwidth", *IEEE, Proc. Microw. Antennas Propag.*, vol. 148, no. 6, pp. 418-422, Dec. 2001.

[30] D. M. Kokotoff, R. B. Waterhouse, and J. T. Aberle, "An annular ring coupled to shorted patch",*IEEE Trans. Antennas Propag.*, vol. 45, pp. 913-914, May 1997.

[31] R. B. Waterhouse, "Broadband stacked shorted patch", *Electron. Lett.*, vol. 35, no. 2, pp. 98-99, Jan. 1999.

[32] Z. Liu, P. Kooi, L. Li, M. Leong, and T. Yeo, "A method for designing broadband microstrip antennas in multilayered planar structures", *IEEE Trans. Antennas Propag.*, vol. 47, pp. 1416-1420, Sep. 1999.

[33] M. A. Matin, B. S. Sharif, and C. C. Tsimenidis, "Probe fed stacked patch antenna for wideband applications", *IEEE Trans. Antennas Propag.*, vol. 55, pp. 2385-2388, Aug. 2007.

[34] K. L. Wong and Y. F. Lin, "Microstrip-line-fed compact broadband circular microstrip antenna with chip-resistor loading", *Microw. Opt. Technol. Lett.*,vol. 17, no. 1, pp. 53-55, Jan. 1998.

[35] K. L. Wong and Y. F. Lin, "Small broadband rectangular microstrip antenna with chip-resistor loading", *Electron. Lett.*, vol. 33, no. 19, pp. 1593-1594, Sep. 1997.

[36] K. L. Wong and J. Y. Wu, "Bandwidth enhancement of circularly polarized microstrip antenna using chip resistor loading", *Electron. Lett.*, vol. 33, no. 21, pp. 1749-1751, Oct. 1997.

[37] C. Y. Huang, J. Y. Wu, and K. L. Wong, "Broadband circularly polarized square microstrip antenna using chip-resistor loading", *Proc. Inst. Elect. Eng. Microw. Antennas Propag.*, vol. 146, no. 1, pp. 94-96, Feb. 1999.

[38] G. S. Hong and S. Li, "Theory and experiment of dual mode microstrip triangular patch resonators and filters", *IEEE Trans. Microw. Theory Tech.*, vol. 45, no.12, pp. 1237-1243, Apr. 2004.

[39] L. Athukorala and D. Budimir, "Compact dual-mode open loop microstrip resonators and filters", *IEEE Microw Wireless Compon. Lett.*, vol. 19, no. 11, pp. 698-700 Sep. 2009.

[40] H. An, B. K. J. C. Nauwelaers, and A. R. V. D. Capelle, "Broadband microstrip antenna design with the simplified real frequency technique", *IEEE Trans. Antennas Propag.*, vol. 42, no. 2, pp. 129-136, Feb. 1994.

[41] B. Froppier, Y. Mahe, E. M. Cruz, and S. Toutain, "Integration of a filtering function in an electromagnetic horn", *in Proc. 33th Eur. Microw. Conf.*, pp 475-478. 1992.

[42] F. Queudet, B. Froppier, Y. Mahe, and S. Toutain, "Study of a leaky waveguide for the design of filtering antennas", *in Proc. 33th Eur. Microw. Conf.*, pp. 943-946. 2003.

[43] N. Yang, C. Caloz, and K. Wu, "Co-designed CPS UWB filter-antenna system", *in Proc. IEEE AP-S Int. Symp.*, pp. 1433-1436. Jun. 2007.

[44] A. Abbaspour-Tamijani, J. Rizk, and G. Rebeiz, "Integration of filters and microstrip antennas", *in Proc. IEEE AP-S Int. Symp.*, pp. 874-877, Jun. 2002.

[45] W. L. Stutzman and G. A. Thiele, *Antenna Theory and Design.* New York: Wiley, 1998.

[46] D. M. Pozar, *Microwave Engineering*, 3rd Ed. New York: Wiley, 2005, ch. 8.

[47] M. Matsuo, H. Yabuki, and M. Makimoto, "The design of a half-wavelength resonator BPF with attenuation poles at desired frequencies", *in IEEE MTT-S Int. Microw. Symp. Dig.*, pp. 1181-1184, 2000.

[48] Yikai Chen, Shiwen Yang, and Zaiping Nie, "Bandwidth Enhancement Method for Low Profile E-Shaped Microstrip Patch Antennas", *IEEE Transactions on antennas and propagation*, Vol. 58, no.7, pp. 2442-2447, July 2010.

[49] Mohamed H. Awida, S. H. Suleiman and A. E. Fathy, "Substrate ?Integrated cavity-backed patch arrays: A low cost approach for bandwidth enhancement", *in IEEE Trans. On Antenna and propg.*, vol. 59, no.4, pp. 1155-1163 Apr. 2011.

[50] J. B. Jadhav., Prof. A. B. Nandgaonkar, "Broadband Stacked patch antenna for PCS and WLAN Application", *in International conference ICAC 2008 Advances in computing*, pp. 39 held at Anuradha coll. Of Engg. Chikhali , Buldhana, India. Feb. 2008.

[51] J. B. Jadhav , A. B. Nandgaonkar, "Design of Broadband Dual polarized Electromagnetically coupled stacked patch for Bluetooth Application", *Proceedings of International conference ICEVD08 Emerging Techniques in Computing , Electronics, Embedded System and VLSI Design*, pp-31, , PDVPPCOE, Ahmednagar, India. March 2008.

[52] L. Athukorala, D. Budimir, "Design of compact dual-mode microstrip filters", *in IEEE Trans. On Microw. Theory Tech.*, vol. 58, no.11, pp. 1155-1163 Nov. 2010.

[53] Z. Jianhong, C. Xinwei, H. Guorui, L. Li, and Z. Wenmei, "An Integrated Approach to RF Antenna-Filter Co-Design", *Antennas and Wireless Propagation Letters, IEEE*, vol. 8, pp. 141-144, 2009.

[54] F. Queudet, I. Pele, B. Froppier, Y. Mahe, and S. Toutain, "Integration of passband filters in patch antennas", *32nd European Microwave Conf., Milan*, pp. 685-688, September 2002.

[55] B. Atallah, M. Jan, and O. Abbas, "Microstrip-CPW bandpass filter for antenna application", *Microwave and Optical Technology Lett.*, vol. 50, no. 1, pp. 51-55, June 2007.

[56] L. Wang-Sang, K. Jin-Hyun, L. Won-Gyu, S. Kwang-Seop, L. Hyung-Sun, and Y. Jong-Won, "Capacitively Coupled Band-Stop Filter with an Integrated Antenna", *IEEE MTT-S, Int. Microwave Symp. Dig.*, pp.2019-2022, June 2006.

[57] C. Pei-Ling, W. Syu-Sheng, and L. Yu-De, "Integrated antennas and diplexers for dual-band wireless local area network (WLAN) system", *Asia-Pacific Microwave Conf., Suzhou, China*, vol. 4 pp. 4-7, Dec. 2005.

[58] D. Veysel, H. Chun-Wen Paul, and E. Atef, "Novel dual-band WLAN antennas with integrated band-select filter for 802.11 a/b/g WLAN radios in portable devices", *Microwave and Optical Technology Lett.*, vol. 49, no. 8, pp. 1868-1872, January 2007.

[59] R. Dehbashi, K. Forooraghi, and Z. Atlasbaf, "A Harmonic-Rejecting Inset-Fed U-Slot Antenna for Rectenna Application", *Sarnoff Symposium, IEEE*, pp. 148, March 2006.

[60] L. Inclan-Sanchez, J. L. Vazquez-Roy, and E. Rajo-Iglesias, "Microstrip patch antenna with compact feed to reduce harmonics", *1st European Conf. Antennas and Propagation , Nice, France*, November 2006.

[61] Manidipa Nath, "Review of filter Technique", *International Journal of Engineering Trends and Technology* Volume3 Issue-3, 2012.

[62] Marjan Mokhtaari, Jens Bornemann, "Microstrip Ultra- Wideband Filter with Flexible Notch Characteristic" , *Wireless Engineering and Technology* Volume. 3, pp. 12-17, 2012.